Animal World

C O N T E N T S

The Animal Kingdom 2
Marvelous Mammals 4
Outrageous Reptiles 6
Spineless Wonders 8
Amphibian Habits 10
Under the Ocean Waves 12
On the Wing 14
Animal Journeys 16
Fearsome Hunters.................... 18
Designed for Defense 20
Animal Architects 22
Animal Babies 24
Animal Courtship 26
Now You See Them 28
Strange but True 30

CREATIVE ● EDUCATION

The Animal Kingdom

Millions of different types of animals inhabit the Earth. Animals are grouped according to features they have in common. All are either vertebrate (they have a backbone) or invertebrate (they don't).

Where do I belong?

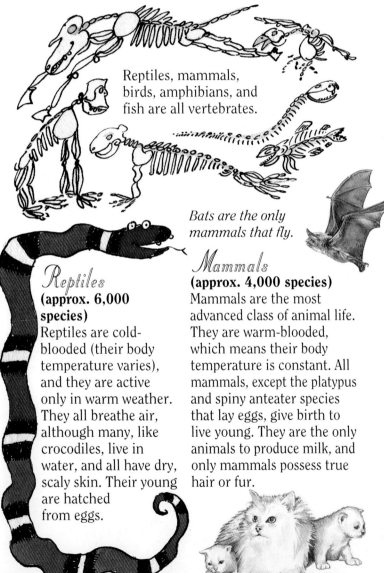

Reptiles, mammals, birds, amphibians, and fish are all vertebrates.

Bats are the only mammals that fly.

Reptiles
(approx. 6,000 species)
Reptiles are cold-blooded (their body temperature varies), and they are active only in warm weather. They all breathe air, although many, like crocodiles, live in water, and all have dry, scaly skin. Their young are hatched from eggs.

Mammals
(approx. 4,000 species)
Mammals are the most advanced class of animal life. They are warm-blooded, which means their body temperature is constant. All mammals, except the platypus and spiny anteater species that lay eggs, give birth to live young. They are the only animals to produce milk, and only mammals possess true hair or fur.

Birds
(approx. 9,000 species)
Most birds can fly. They have streamlined, feathered bodies and wings. Their legs and feet are scaly, and they have beaks for feeding, but no teeth. Birds are warm-blooded, and their young hatch from eggs.

Amphibians
(approx. 3,000 species)
Amphibians are born in water and live the early part of their life there. Later they move to land. They are cold-blooded and lay jelly-covered eggs called spawn.

Fish
(approx. 21,000 species)
Fish are cold-blooded and live in water. They start life as eggs. They have scaly skins, and fins help them to swim. They breathe oxygen from the water through gills.

Invertebrates
Insects are the largest of the invertebrates. There are more beetles than any other insects.

Insects (approx. 1 million species)

Sponges (approx. 5,000 species)

Crustaceans (approx. 25,000 species)

Spiders (approx. 30,000 species)

Mollusks (approx. 40,000 species)

Marvelous Mammals

What does this odd bunch of animals have in common? They are all mammals because of three features they share: they feed on their mother's milk as babies, they all have some hair or fur, and they are all warm-blooded—that means the temperature of their blood stays the same, no matter what the outside temperature is. One flies, several swim, but most species live on land.

Brainy bunch

This is where I fit!

Primates are the cleverest animals—they have the biggest brains. Humans are primates, so are 180 other species, including apes, monkeys, lemurs, and bush babies. Primates have nails, rather than claws, and almost all of them can grasp objects with their hands and feet.

Ocean dwellers

Dolphins, porpoises, narwhals, and other whales are all sea-dwelling mammals, but they need to surface to breathe. The sperm whale dives to 1,000 ft. (300m) below the ocean's surface and can stay under water for almost two hours. Seals and walruses hunt in the water, but breed on land.

Marvelous Mammals

Hair types

Believe it or not these are all different types of hair

WOOL

QUILLS

BRISTLES

FUR

HORN

Mystery mammal
Which mammal has ...

webbed feet?   lays eggs?   and a bill like a duck's?

(Answer at bottom of page).

Gnawing mammals
The common feature of rodents is their four chisel-shaped front teeth. These grow throughout the animal's life, but are kept the same length by continual gnawing. There are 32 families of rodents divided into three main groups: squirrels, porcupines, and rats and mice.

Baby in a pouch
Animals with pouches are called marsupials. They include kangaroos, koalas, and opossums. A baby marsupial is less developed than other mammal babies, so it spends its early days in the pouch, where the mother's nipples are located and it can feed in safety.

DUCKBILL PLATYPUS

Outrageous Reptiles

Reptiles have lived on Earth for more than 300 million years. They evolved from some of the earliest amphibians. There are about 6,000 different types of reptiles, ranging from tiny lizards just a few inches long to snakes more than 33 ft. (10m) long.

PYTHON

Hard to swallow
Snakes always eat their prey whole (they can't chew) and mostly alive. Pythons have been known to eat animals as large as leopards and deer.

Surprise eggs
Alligators, crocodiles, turtles, and a number of snakes are hatched from eggs. So were dinosaurs!

All change!
Several times each year snakes shed their old skin. A brand-new one has grown underneath.

Outrageous Reptiles

Would a blindfolded snake dance for a snake charmer?
No, it has no outside ears and cannot hear airborne sounds, but it is sensitive to vibration through the ground. Its dance actually follows the movement of the snake charmer.

What distinguishes reptiles from other animals?
They are cold-blooded, meaning that their body temperature is about the same as their surroundings. They breathe by means of lungs, they have backbones, and they have dry, scaly skin. Many are hatched from eggs.

LIZARD

TURTLE

Are snakes slimy?
No. They have scales that cover their body which makes them look shiny and wet, but in fact they are dry when you touch them.

I was a reptile once.

Reptile records

- The Nile crocodile can run 35 mph (56kmh).
- The chameleon can shoot out its coiled tongue further than the length of its own body to catch its prey.
- The largest lizard, a komodo dragon, can grow up to 10 ft. (3m) in length.
- The oldest reptile fossil is more than 300 million years old.

CROCODILE

Spineless Wonders

None of the animals on this page has a backbone. Scientifically speaking they are all invertebrates. Here is a selection of the animals that fall into this class.

What is an insect?

Not all creepy-crawlies are insects, though most are. The sure way to recognize an insect is to count its legs. If it has six it is an insect. Their bodies have three parts—the head, thorax, and abdomen. They also have feelers or antenna and most have one or two pairs of wings.

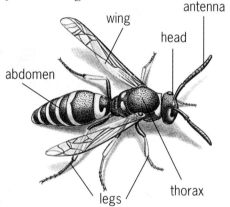

Since spiders have eight legs and only two body parts they are not insects.

Which one isn't an insect?

BEETLE

BUMBLEBEE

GRASSHOPPER

SPIDER

BUTTERFLY

(*The answer is on this page.*)

Squidgy bodies

The soft-bodied creatures of the animal kingdom such as snails, mussels, oysters, and squids, are known as mollusks. Most of them have a hard shell to protect their body, but some, like octopuses and certain slugs, have nothing.

Spineless Wonders

In the ocean
Believe it or not, everything on this page is actually an invertebrate animal. The jellyfish 1. looks like a floating umbrella and has a poisonous sting. The anemone 2. may look like a flower, but its "petals" are deadly tentacles that paralyze passing fish. The anemone then eats them. The coral 3. seems to do nothing, but it eats stray bits of food floating in the water; so does the sponge. 4. The crab 5. is armor-plated, and the squid 6. can hide itself from predators by squirting a cloud of ink into the water. The clam 7. opens its wavy shell to feed.

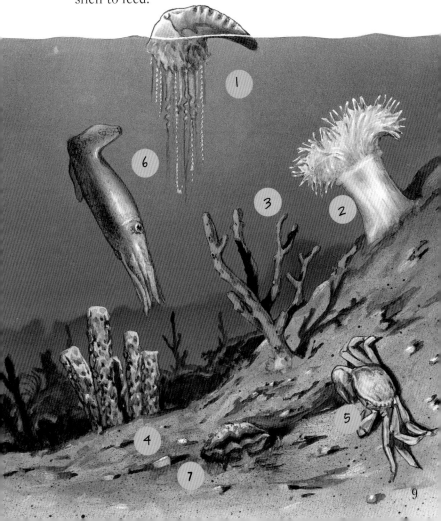
STARFISH

Amphibian Habits

Amphibians are animals that are born in water and spend the first part of their life there. Later they move to land. Frogs and toads are in one family, salamanders another, but there is a third less common group called caecilians: wormlike creatures from the tropics.

Heron fodder
Frogs lay thousands of eggs. Even if the tadpoles escape their many underwater predators, more are waiting on land, like this heron.

2 weeks after the eggs are laid the tadpoles emerge.

6 weeks later the back legs appear.

10 weeks later the front legs appear.

12 weeks later the tail disappears and the change to a frog is almost complete.

After 2–3 years the frog will go back to the water to mate. If it isn't eaten it could live for 10 years.

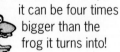

Big baby little frog!

The paradoxical frog is so-called because as a tadpole it can be four times bigger than the frog it turns into!

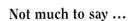

Not much to say …
Scientists don't know much about caecilians. They look like overgrown worms—many adults become blind and live underground. They are shiny, hard to hold, and boring!

Lizard look-alikes
Salamanders are often mistaken for lizards, but they are different. Lizards have scales and claws, salamanders do not. Not all salamanders are small. The giant salamander of Japan can grow up to 5 ft. (1.5m) long.

A strange tail
If attacked, some salamanders can leave their tails behind! The tail continues thrashing after it breaks off, which distracts the attacker and gives the salamander time to escape. It later grows a new tail.

Noxious newts
Newts may look harmless, but they are definitely not defenseless. Certain species of newt can ooze a highly poisonous fluid from their skin when attacked.

WARTY NEWT

What makes me a toad?
Toads are related to frogs, but are different. They have dry, warty skin. Frogs have smooth, moist skin. The poor toad can't jump as far as a frog, either!

Under the Ocean Waves

Seas and oceans cover almost three-fourths of the world's surface, and are home to more than 20,000 different kinds of fish. The biggest animal in the ocean is the blue whale, which, like dolphins and seals, is a mammal. The other sea creatures are invertebrates (animals without backbones), such as jellyfish, starfish, crabs, lobsters, shellfish, and worms.

Floating colony
Not to be messed with, the Portuguese man-of-war trails tentacles 65–100 ft. (20-30m) long under the surface. Their sting can kill a human. This jellyfish is actually a colony of tiny creatures drifting across the oceans like a mobile home.

BEFORE...

...AFTER

Prickly customer
The porcupine fish is bigger than a predator might think. At the first sign of danger it swallows a large quantity of water or air and blows itself up into a ball with stiff spines sticking out all over it.

Safety in numbers
Up to 50 lobsters might travel in a train like this. If attacked, they form a circle with their pincers facing outward, like a Wild West wagon train.

Under the Ocean Waves

Bad table manners
Sharks mostly eat other fish, but about 100 people are attacked by them each year, of whom about 50 die. Here are some other things that have been found inside sharks:

Great white sharks can grow to 35 ft. (11m) in length.

FUR COAT   PORCUPINE   DOG'S HEAD

Razor-sharp
The sawfish not only uses its sawlike snout to stir up the seabed in search of food, but if it passes a school of fish, it uses it to impale its prey.

SAWFISH

The gulper
There isn't much food in the ocean depths, and this mean-looking beast has developed huge jaws so that it can eat fish much bigger than itself.

GULPER

Ocean depths

At its deepest the ocean floor is almost 7 miles (11km) from the surface, and below 2,000 ft. (600m) the water is permanently dark, so many fish carry their own lights.

On the Wing

Only birds, bats, and insects truly fly. But certain mammals, reptiles, and even fish seem to fly. Some birds, such as ostriches, have wings but can't fly.

HUMMINGBIRD

Designed for flight
The number of feathers on a bird ranges between about 1,000 for some species of hummingbird to more than 25,000 for large birds such as swans. Birds' feathers always weigh more than their hollow skeletons.

Down is found close to the skin and provides insulation.

Flight feathers are longer and much stiffer.

Contour feathers give a streamlined surface to the bird's body.

Different wings for different jobs

Albatross wings are good for gliding.

The swift's wing shape is designed for speed.

The shape of the kestrel's wings allow it to hover.

Wings don't just flap up and down, they also move backward and forward to provide lift.

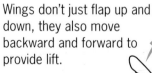

On the Wing

Acrobats!
Bats' wings are actually made of skin stretched from their bodies to the bones of their front and back legs. When they fly, they look as if they are doing the breaststroke in the air. These furry creatures are nocturnal, and they sleep hanging upside down.

Dazzling dragonfly
This beautiful insect spends the first year of its life as a dull-looking, wingless nymph, crawling about underwater.

Champion gliders

The South American flying frog uses the membranes between its toes to turn its jumps into glides.

The sugar glider of Australia can glide for up to 1,500 ft. (450m).

The flying lizard uses its spiny flaps to fly.

Animal Journeys

Many animals make journeys every day, looking for shelter, food, or water. Some, however, make enormous journeys seasonally.

The Arctic tern travels farthest, more than 10,850 miles (17,500km) twice a year.

Champion travelers
Birds make the longest regular journeys of all animals. Swallows, geese, ducks, cranes, and even small birds such as thrushes and starlings travel hundreds or thousands of miles.

The long march
As the Antarctic winter closes in and ice spreads farther out from land, emperor penguins must travel farther to feed in the sea. They cover up to 90 miles (144km) in 10 days. The temperature can drop as low as –49°F (–45°C) with winds of up to 125 mph (200kmh).

EMPEROR PENGUINS

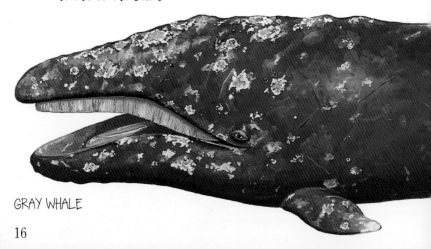

GRAY WHALE

Animal Journeys

Monarchs of the air
Each fall, clouds of monarch butterflies migrate to California, Florida, and Mexico. Some fly 2,000 miles (3,200km) from Canada and the northern U.S. Every now and then some are blown across the Atlantic Ocean to Europe.

MONARCH BUTTERFLIES

CARIBOU

Caribou caravan
Caribou are found in North America and are the same species as the European reindeer. Both reindeer and caribou travel south in the fall and north in the spring.

No phone needed?

Whales travel huge distances to feed and give birth to their young. Many sing to each other while traveling. Some whale songs can be heard 1,550 miles (2,500km) away from the singer.

Fearsome Hunters

Animals that hunt other animals are often armed with such deadly weapons as sharp claws and teeth. Other animals use stealth, creeping up on their victims, while still others use chemical weapons—poison—to stun or kill.

Powerful pussycats

The big cats are powerful. Lions and jaguars can kill with one swipe of their paw or a bite to the neck. The Siberian tiger can jump 16.5 ft. (5m) from a stationary position.

Danger behind the door

The trap-door spider makes a camouflaged door for its nest and lies in wait under it.

When a victim moves into range the spider pounces. It drags the victim into the nest, closing the trap door behind it.

Paralyzing the victim with a bite and hidden from danger, the spider can eat its meal in safety.

Fearsome Hunters

Gone fishing
The fish eagle skims low over the water, catching fish in its talons. The fish-eating bat uses the same trick, and has a long forward-facing claw on each foot which it trails through the water.

With a fish in its talons the eagle heads for a safe perch to eat its catch.

Eight-legged killers
Many spiders use poison to paralyze their prey. The poison is injected through hollow fangs. Large tarantulas will catch mice, lizards, and even birds. The fishing spider rests on the surface of the water and seizes insects and even small fish that swim below the surface.

BIRD-EATING SPIDER

Bull's-eye!
The archerfish is a crack shot. It spits at insects that settle near the water, knocking them onto the surface where it can snap them up.

Ancient hunters
Sharks are such perfect hunters that they have remained basically unchanged for millions of years. Their sense of smell is so acute that they can detect one drop of blood in one million drops of seawater.

Designed for Defense

Many animals try to defend themselves against attack from predators. Often this is by the use of camouflage, or disguise. Other methods include armor, chemicals, and poison.

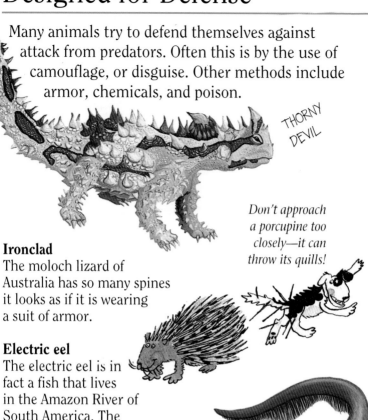

THORNY DEVIL

Ironclad
The moloch lizard of Australia has so many spines it looks as if it is wearing a suit of armor.

Don't approach a porcupine too closely—it can throw its quills!

Electric eel
The electric eel is in fact a fish that lives in the Amazon River of South America. The electricity it produces is so powerful that it can easily knock over a horse. The electric eel has no natural predators.

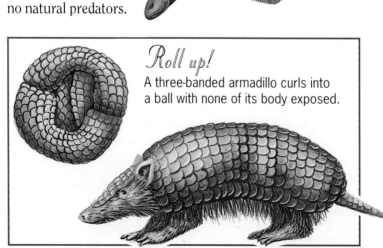

Roll up!
A three-banded armadillo curls into a ball with none of its body exposed.

Designed for Defense

Warning—poisonous
Like many poisonous creatures, poison-arrow frogs of South America advertise their danger to others with bright colors. The most poisonous, the golden poison-arrow frog, contains enough poison to kill 1,500 humans!

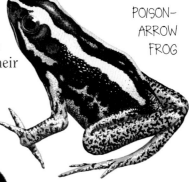
POISON-ARROW FROG

Phew!
- Skunks spray an offensive smell that can be detected more than a third of a mile (half a kilometer) away.

- Shrews have stink glands that deter other animals from eating them—they taste disgusting!

- Stink bugs eject a foul-smelling liquid when in danger. One species can squirt its spray up to a foot (30cm).

All puffed up
The chuckwalla lizard has an unusual form of defense. It darts into a crevice in the rocks and then inflates itself until it becomes firmly wedged into the crevice and cannot be pulled out!

Hagfish mucus
Some deep-sea hagfish can grow to 6.5 ft. (2m). At the first sign of trouble a hagfish produces large quantities of mucus, turning the surrounding water slimy and making gripping the fish very tricky.

Playing possum
Many animals hunt only live animals. So, when all else fails, why not play dead? This is what the opossum does, hence the expression "playing possum."

Animal Architects

Not all animals build homes, and some make very crude ones. On this page you can see the more ambitious ones built by some of the best animal architects!

Weaverbirds
African weaverbirds are famous for their elaborate nests. Some build communal nests in the crowns of trees that are larger than any other birds' nests. Others build individual nests like hanging fruit. On the left you can see just some of the clever examples of their weaving.

Beavering away
The busy beaver earns its name the hard way! A whole family, or several families, may join in building a dam that can completely block a river. The beavers enter their new home from the lake that forms behind it. They think nothing of felling whole trees.

Animal Architects

House of silk
Spiders use their own silk mills to weave intricate homes, which unfortunately need constant repair. Only the spider knows which are the sticky strands designed to catch prey, while it moves along the dry ones.

Best nests

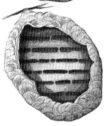

This South American wasp's nest is built of clay and sand and will last as a home for years.

A bee's honeycombed hive is a perfect geometrical formation of hexagons. The walls are made of wax.

Termite tower
Lots of tiny termites built this amazing pagodalike structure. The series of roofs protect the termites from the torrential rains of a South American rain forest. The cutaway shows the air-conditioned chambers in which the termites live.

Animal Babies

Some animals, such as mammals, take great care of their young, feeding and protecting them during their early life. Other animals, such as insects, leave their young to fend for themselves.

Protective penguins
The emperor penguin of Antarctica holds its egg on its feet, tucked under a flap of feathers and skin to stop the cold from freezing the egg solid. Once the baby is hatched, it will stay close to its parents to keep warm.

What a mouthful
The mouthbreeder is unusual among fish in taking some care over its young. The little fish stay close by their parent and dart into the safety of its mouth if a predator approaches.

How many?
Larger animals tend to have few babies at one time, but smaller or more primitive animals, such as sponges, produce thousands or even millions. The giant clam produces one billion eggs each year.

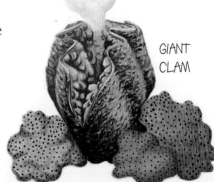

GIANT CLAM

Animal Babies

The biggest baby in the world

The biggest baby is the calf of the blue whale. It is about 23 ft. (7m) long when born and weighs 2 tons. It drinks up to 100 gallons (3.8 liters) of its mother's milk each day and puts on weight at the rate of more than 200 lb. (90kg) per day during the first year of its life.

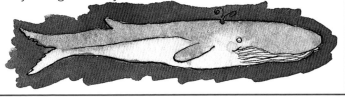

At a signal from their mother, young shrews form a line. Each holds the tail of the one in front in its mouth as they head for safety.

Babies on board
The marsupial frog puts its eggs in a pouch on its back. The eggs hatch into tadpoles which grow into little frogs, all in the safety of the pouch.

Kangaroos
The joeys stay in the pouch until they are old enough to fend for themselves.

When the young frogs are old enough, their mother opens the pouch to let them out.

Animal Courtship

Many animals use color, sound, smell, and even light when trying to attract a mate. Others perform complicated dance routines or build special displays.

Who says I'm a showoff?

Perfume trail
The female peppered moth produces a scent so strong that she can attract males in an area of up to 24,000 square yards (20,000sq.m).

Bower builder
The male bowerbird builds a small chamber which he decorates with berries, shells, flowers, or feathers. He then dances at the entrance to attract a female mate into his bower.

Fantastic feathers
When he wants to impress a female mate, the male riflebird lifts his wing feathers into an amazing arc and opens his beak to show off his bright yellow mouth.

Sing a song
Mole crickets build a burrow that works like a megaphone. Their song can reach 150 decibels and can be heard over 1,650 ft. (500m) away.

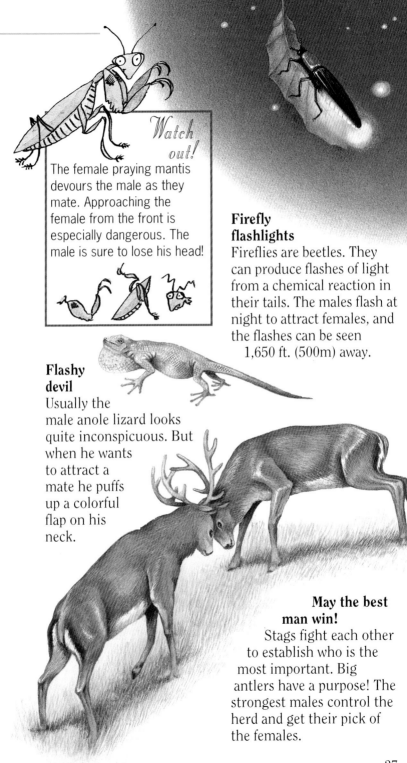

Watch out!

The female praying mantis devours the male as they mate. Approaching the female from the front is especially dangerous. The male is sure to lose his head!

Firefly flashlights

Fireflies are beetles. They can produce flashes of light from a chemical reaction in their tails. The males flash at night to attract females, and the flashes can be seen 1,650 ft. (500m) away.

Flashy devil

Usually the male anole lizard looks quite inconspicuous. But when he wants to attract a mate he puffs up a colorful flap on his neck.

May the best man win!

Stags fight each other to establish who is the most important. Big antlers have a purpose! The strongest males control the herd and get their pick of the females.

Now You See Them...

Confusing the enemy is the name of this game. Animal skins are often patterned for a purpose. Here are some tricks aimed at disguising the hunter and the hunted.

Northern stoats turn white in winter.

Farther south they are white and brown.

This jaguar is hiding in a tree, waiting for prey to pass underneath.

All change
The stoat's fur changes color according to the season and how far north it lives. Southern stoats do not change color in the winter at all.

This trumpet fish is hiding in a school of fish. It often does this to enable it to approach prey unnoticed.

Now You See Them ...

Which one is poisonous?

You don't know—nor does a predator! The poisonous red-cheeked salamander doesn't get eaten. By developing similar red cheeks, the harmless dusky salamander hopes to avoid being eaten too.

... Now you don't!

Many creatures disguise themselves to avoid being eaten by predators. Each of these animals is imitating something else. What do you think they are imitating? (Sometimes there are clues in their names.)

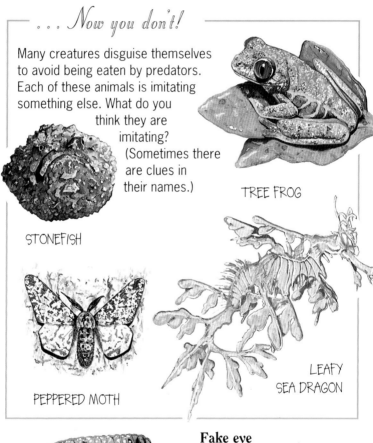

STONEFISH

TREE FROG

PEPPERED MOTH

LEAFY SEA DRAGON

Fake eye

The cometfish has an "eye" near its tail. If a predator strikes there the cometfish is in less danger than if it struck its real eye. Many animals use this kind of camouflage.

Strange but True

Going to great lengths
The longest snake in the world is the reticulated python. It can exceed 21 ft. (6.3m). One of these snakes, kept in a zoo, managed to swallow a pig weighing 120 lb. (54kg). A reticulated python has been known to go without food for 570 days, so no wonder it eats big meals!

Wake up!
People say that the Dall's porpoise never sleeps at all. On the other hand, some armadillos spend four-fifths of their lives sleeping.

Minute mammal
The bumblebee bat is only 1 in. (2.5cm) long and has a wingspan of just 6 in. (15cm).

Enormous eye
The giant squid has the largest eyes in the world: they can be up to 16 in. (40cm) in diameter!

What a racket
Howler monkeys are the noisiest animals in the world. They can be heard 10 miles (16km) away.

Bumper jumper
Fleas are the best jumpers of all animals. They can leap more than a hundred times their own height.

Jumbo egg
Ostriches lay the biggest eggs in the world. They lay six to eight eggs and each one has a volume equivalent to 25 chicken eggs.

Skyscraper nest
A pair of bald eagles built a nest 9.6 ft. (2.9m) wide and 20 ft. (6m) deep. It was examined in 1963 and estimated to weigh more than 2 tons.

Cheetahs can run 60 mph (96kmh).

Spidery facts

This Chilean red-legged spider has the hairiest legs. But the goliath bird-eating spider has the longest legs: One was found in Venezuela with legs that were 11 in. (28cm) long.

Long birth, short life
The nymph of the mayfly may live in a pond for years, but when it emerges from the water as an adult it lives for just a few hours!

The oldest lizard
A male slow-worm kept in the Copenhagen zoo lived for 54 years. Other than that it was not very remarkable.

Poison sting
The Palestine scorpion kills its prey with just a drop of the deadliest venom in the world.

Irritating insect
For self-defense, the blister beetle oozes a highly itchy substance from its jointed legs to repel any predators from eating it.

Piranha peril

The razor-toothed piranhas of South America are the most ferocious freshwater fish in the world. They feed in schools, sometimes numbering thousands. On September 19, 1981, more than 300 people were eaten by them when a boat capsized in Obidos, Brazil.

Ancient giants
Giant tortoises are believed to live nearly 200 years.

Albatross 14	Narwhals 4
Alligators 6, 7	Nests 22, 23, 30
Amphibians 3, 10-11	Newts 11
Anemone 9	Octopuses 8
Apes 4	Opossums 5, 21
Armadillo 20	Ostriches 14, 30
Bats 2, 15, 19, 30	Oysters 8
Beavers 21	Penguins 24
Beehive 23	Porcupines 5, 13, 20
Birds 3, 14-15, 19, 21	Porpoises 4, 30
Caecilians 10, 11	Primates 4
Clam 9, 24	Pythons 6, 30
Coral 9	Rats 5
Crabs 9, 12	Reptiles 2, 6-7
Crocodiles 2, 6, 7	Rodents 5
Crustaceans 3	Salamanders 10, 11, 29
Dinosaurs 6	Scorpions 3, 8, 31
Dolphins 4, 12	Seals 12
Dragonflies 15	Sharks 13, 19
Duckbill platypus 5	Shrews 21, 25
Electric eels 20	Skunks 21
Fish 3, 12, 13, 19, 21, 24, 28, 29, 31	Slugs 8
Fleas 30	Snails 8
Frogs 10, 11, 15, 21, 25, 29	Snakes 6, 7, 30
Humans 4	Spiders 3, 8, 18, 19, 21, 23, 31
Insects 3, 8, 14, 15, 24, 31	Sponges 3, 9, 24
Invertebrates 2, 3, 8-9	Squids 8, 9, 30
Jaguars 18, 28	Squirrels 5
Jellyfish 9, 12	Starfish 9, 12
Kangaroos 5, 25	Stink bugs 21
Koalas 5	Stoats 28
Lions 18	Tadpoles 10
Lizards 6, 7, 11, 15, 20, 21, 31	Tarantulas 19
Lobsters 12	Termite tower 23
Mammals 2, 4-5, 24	Tigers 18
Marsupials 5, 25	Toads 10, 11
Mayflies 31	Tortoises 31
Mice 5	Turtles 6, 7
Mollusks 3	Vertebrates 2
Monkeys 4, 30	Whales 4, 12, 25
Mussels 8	Wings 14-15

Published in 1997 by Creative Education
123 South Broad Street
Mankato, Minnesota 56001
Creative Education is an imprint
of The Creative Company
Cover design by Eric Madsen
Illustrations by Jon Evans, Charlotte Hard, Ron
Hayward, Sharon McCausland, Deborah Pulley

Text © HarperCollins Publishers Ltd. 1996
Illustrations © HarperCollins Publishers Ltd. 1996
Published by arrangement
with HarperCollins Publishers Ltd.
International copyrights reserved in all countries.
No part of this book may be reproduced in any form
without written permission from the publisher.
Printed and bound in Hong Kong.

**Library of Congress
Cataloging-in-Publication Data**
Animal world.
p. cm. - (It's a fact!)
Includes index.
Summary: Presents miscellaneous facts about
animals all over the world.
ISBN 0-88682-856-2

1. Animals-Juvenile literature. [1. Animals-
Miscellanea.] I. Creative Education, Inc.
(Mankato, Minn.) II. Series.
QL49.A585 1997   96-35292
591-dc20
EDCBA